The Flower Shop

by Richard Kehl & Cooper Edens

Blue Lantern Books

If you walk into the Town of Shops early on a summer morning you will find the residents just opening their street windows or their carved front doors. The splash of fountains will be the loudest sound. The many small gardens will be glowing, and the leaves of the trees you pass under will still be moist. The shops will all be closed at this hour, but if you cross a little green bridge that offers itself just past the town's central square you will find The Flower Shop. The lady who keeps it will be outside watering and arranging the flowers and plants which grow in it's window boxes and border the entrance. If you ask, she will lead you inside. The confusion of color and odors will make you glad. She will lead you from room to room, revealing floral treasures of great rarity and variety. Here in these pages we show you some of the things we were shown on our many visits. We gather them here in remembrance of the visits made and the gifts secured.

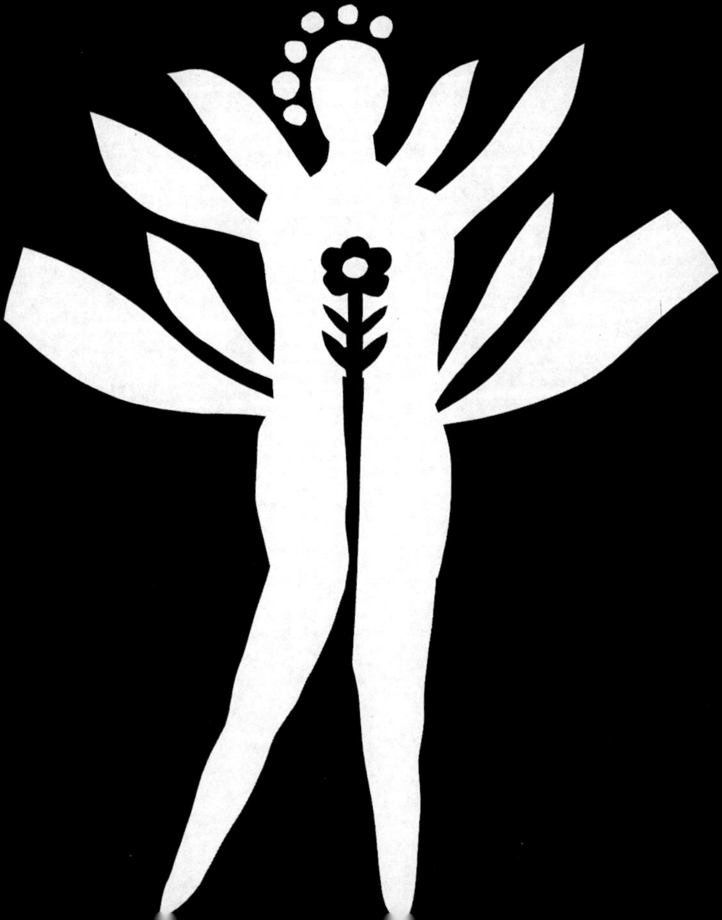

Sipping the cold soup made from the chrysanthemums of dreams.

Paul Carroll

Though modest and small, and without much color, a blossom has all the qualities of an explosion.

Colette

I perhaps owe having become a painter to flowers.

Claude Monet

Spring unlocks flowers to paint the laughing soil.

Reginald Heber

Lavender's blue, dilly, dilly, lavender's green. When I am king, dilly, dilly, you shall be queen.

Author unknown

The career of flowers differs from ours only in audibleness. I feel more reverence for these mute creatures whose suspense or transport may surpass ours.

Emily Dickinson

Time to plant tears, says the almanac.

Elizabeth Bishop

A garden I tend whose blossoms never existed.

Pablo Neruda

A lot of good it does to define, name, or forsee what ignorance lets me look upon as wondrous! Neither the flower nor it's influence can be explained.

Colette

Flowers are words which even a babe may understand.

Bishop Coxe

A woman tries to comfort me. She puts her hand under my shirt and writes the names of flowers on my back.

Mark Strand

The blossoms are so short, the forgetting so long.

Walt Whitman

When I have arranged a bouquet, in order to paint it, I go around to the side that I have not looked at.

Pierre-Auguste Renoir

Send forget-me-nots anonymously.

Author unknown

More things grow in the garden than the gardener sows.

Spanish Proverb

SPLENDESCAM, DA MATERIAM...

30 ans ou la vie en rose
Raoul Dufy

Le Nez
Aquilin:
Esprit, Perspicacité
Droit:
Distinction
Long et poin
Curiosité
Relevé:
Jovialité, Gaité.
Fort:
Bonté, Faiblesse

D
C

Picture Credits

Front cover. Richard Kehl, collage.
Endpapers. Richard Kehl, collage.
Frontispiece. Anonymous,Turkish painting of Mohammed II, 15th Century.
Facing preface. Anonymous, scrap, late 19th Century.
2. Richard Kehl, collage. **3.** Cooper Edens, collage.
4/5. Cooper Edens, collage. **6.** Richard Kehl, collage.
7. Cooper Edens, collage. **8.** Anonymous, circa 1807.
9. Giuseppe Arcimboldo, 1573.
10/11. Charles Courtney Curran, *Lotus, Lilies,* 1888.
12. Artist unknown, stone lithograph, circa 1900. **13.** A Japanese tattoo.
14. Martin Johnson Heade, 1863. **15.** Richard Kehl, collage.
16. Jaques Henri Lartigue. **18/19.** Anonymous, Dutch tulip beds.
21. Carl Larsson, *Azalea*, 1906. **22.** Richard Kehl, collage.
23. Richard Kehl, collage. **24.** Raoul Duffy, 1931.
25. Severin Rosen, detail, 1848. **26.** Richard Kehl, collage.
27. Johann Jensen, *A Basket of Flowers.* **28/29.** Postcard, Rose City Festival, 1910.
30. Anonymous, English textile design, 19th Century.
31. Anonymous, *The Nose*, French Postcard.
32. Anonymous, poster, 1910.
33. George Frederick Watts, Portrait of Dame Alice Ellen Terry.
34/35. Richard Kehl, collage. **36.** Debora Greger, collage, 1984.
37. Anonymous. **39.** Richard Kehl, collage.
Back cover. M. Hart, *Persian Sun's Eye Tulip*, 1828.

This book was designed and set in Melior by
The Blue Lantern Studio.

Printed in Hong Kong

ISBN 0-9621131-2-3